THE EFFECTIVE INSPECTION REPORT

THE ART OF WRITING PERCUSSIVE INSPECTION REPORTS

Published by Ludovic Gaillard

Collection QCLeaks

Ludovic GAILLARD

Copyright 2017 Ludovic Gaillard

Introduction --5

Section 1: The 5 mistakes to avoid---8
Error 1: the "summary" report ---9
Error 2: the "saga" way --10
Error 3: Approximations--10
Error 4: The dangers of "copy / paste" ----------------------------------12
Error 5: The off topic --12
Conclusion --14

Section 2: A Complete Report ---16
Cover Page--17
The plan --20
   1- List of reference documents --20
   2- Visual and dimensional inspection----------------------------------21
   3- Non-destructive testing---22
   4- Mechanical test and pressure test ----------------------------------24
   5- Functional tests---25
   6- Packaging and preservation --26
   7- The documentary review ---27
Photos and diagrams---28
Conclusions--29
The use of an imposed form ---30
Conclusion --31

Section 3: Small---32
Comment on the progress---32
Safety commentary---35
Additional information ---37
The Flash message --38
Conclusion --39

Section 4: The punch list ---41
The form ---42
Photographs and other media ---43

Responsibilities -- 44
Managing the punch list -- 45
Conclusion -- 46

Section 5: Attachments -- 47
Photos -- 47
Reports and certificates --- 49
Annotated Plans -- 50
Others (video, files) -- 50
Referencing attachments --- 51
Distribution form of attachments --------------------------------------- 52
Conclusion -- 53

Section 6: save time -- 54
Write on the spot -- 54
Annotated documents --- 55
The use of "copy / paste" --- 56
Use of reference documents -- 57
Conclusion -- 58

Section 7: Distribution of the report --------------------------- 59
The format of your report --- 59
Distribution to the customer --- 61
The "Flash" report --- 62
Distribution to the VENDOR -- 63
The Signatures --- 64
Conclusion -- 65

Last conclusion --- 67

Notice and Contact -- 69

Thanks --- 70

INTRODUCTION

If the inspection can be considered as an art, the writing of an inspection report is also an art. Many inspectors, however competent they may be, are undervalued because their reports have flaws or shortcomings that their recipients considered as objectionable or unacceptable. Too much information or not enough, approximations, but also typographical errors such as copy / paste unfortunate, such are the imperfections, that as a Quality Control Coordinator, I have observed in 20 years of career.

Do you want to improve your reputation without having to (too much) change your work habits? Make your reports indispensable documents for your clients? Do you want your reports to be references for other novice inspectors or not?

This manual is intended for all inspectors, whether you are an independent company or a part of a Quality Control company. This guide is also useful for all those who are in charge of inspection reports and in charge of their operations such as Quality Control Managers or coordinators.

This manual is not intended to teach you the inspection. It aims to bring you these few keys that will enhance your relationship with simple methods that will allow you to be recognized not only as an inspector, but also as an editor.

For my part, I have been using these methods since I started, so I can confirm its effectiveness in petrochemicals. I had the opportunity to confront this exercise both as an inspector and as a quality control coordinator and I had the opportunity to train many inspectors in this method. I am practicing all these activities as I write this manual. I therefore propose to share with you my method which has proved its effectiveness in all circumstances. You can, of course, apply some or all of the techniques I am giving you here.

In order to cover all aspects of the inspection report, I suggest that you proceed in stages. Each of these sections will address different themes that are described below.

Section 1 will identify errors to be avoided. I myself have been able to observe these mistakes in the exercise of my profession and even unfortunately sometimes committed.

Section 2 will focus on the content of the inspection report, detailing the plan. This plan will be the skeleton of your report and the basis of your writing. I will also give you some editorial advice sometimes also inspection.

Section 3 will deal with the "punch list" or task list. Form and responsibility are topics that will be addressed as well as the basics of managing this document. The subject being vast, it will be the subject of another manual which will go into the details of this particular document.

Section 4 will address the topic of attachments. Your report will most certainly be accompanied by documents that you will need to manage properly to maintain the clarity you are looking for.

In Section 5, I will share with you these "little extras" that will make your report a must-have document and help set you apart from other inspectors.

With section 6, we will review some methods that will save you time in writing your report.

Section 7 is about disseminating your report to both your customer and the supplier you inspected.

Finally, we will conclude this manual with a short summary as well as my direct contact details, because I am interested in your experiences and comments. Remember that this collection is yours and I will remain attentive to your remarks to improve each of the volumes that will be published or updated.

SECTION 1: THE 5 MISTAKES TO AVOID

As with any exercise, there are pitfalls to avoid. This is even more true for an inspection report whose subject does not suffer from any approximation.

The pitfalls are many and it is important for the good behavior of your report to be careful not to be victimized. Rest assured, it is quite feasible and easy enough if you pay a little attention. Of course, with experience, you will not be trapped anymore and you will write your report without worrying about it. I myself have fallen into some of these and I thank my mentors for giving me the right path.

A golden rule that I will often repeat in this book: the report you write is not intended for you. Remember that you are the eyes and ears of your clients and, in this sense, they rely on you to demonstrate the precision and effectiveness that are legitimately expected.

In this step, I deliberately limited the number of traps to those commonly encountered so as not to overload this manual. But if you can already avoid these errors, you will only have to hunt the final details that will be specific to the inspection you have carried out.

ERROR 1: THE "SUMMARY" REPORT

Over the course of my career, I have too often had the opportunity to see reports on my desk too concise to be exploitable. They were more like abstracts - even even shorter ones: 3 lines with no precision or really information, with meanings difficult to understand and a conclusion as restricted as possible: "accepted", "accepted with comments ",...

What would you do with this type of report? Nothing, of course. Not even if you are the editor and you have to resume it a few weeks, months or years later.

Once again, your sponsors, whether internal or external, are not present during the inspection, so they need concrete and clear details to get a sense of what you have been able to control. It is necessary for the various recipients of your report to understand everything that has happened and what can result from it.

With a well organized plan, you can avoid this error and deliver a report that is accurate enough to be usable. Remember that when you report problems, your clients must be able to respond.

Beware, however, of not falling into excess ... Error number 2.

ERROR 2: THE "SAGA" WAY

In contrast to the "summary" report we find the "saga" type report.

It has happened to me also to receive real novels perfectly indigestible. The reports transmitted were merely interminable sequences of paragraphs which made as much of the inspector's adventures from his residence to his return as the most insignificant and useless details of the inspection.

Think of your sponsors who will have to read, understand and use your report. Also think about the time you would spend writing this type of novel. Unfortunately, there could be consequences of losing your reader in unnecessary detail. Especially if you do not highlight the most important points. Your client may miss something crucial.

Again, it is easy not to fall into this kind of wrong thanks to a strict plan and special attention to the search for relevant information. But, of course, writing presents its own risks ... The trap number 3.

ERROR 3: APPROXIMATIONS

Accuracy is probably one of the most important things in inspection report. What is more vague than to see notions appear such as ugly, taller, greenish, roughly ... or any other expression of everyday language.

Wanting to put details in the inspection report is important, but in addition to being useful, they need to be specific. For example, if you need to compare 2 diameters, indicate the exact size of each diameter rather than saying that one is bigger than the other.

Be sure to indicate measurements - size, pressure, volume, etc. - in the units used in your reference documents. For example, if your pressure test procedure indicates values in bar, do not write your report in psi. Your customer is accustomed to working with these units and therefore, so will not have to make conversion sometimes tedious.

Once again the inspection report is not intended for you. It is very detrimental to write a report by referring to personal notes or to omit information that know from experience or flow of source.

As you will not be the user of this report, the information must be clear and complete. They must be directly exploitable to prevent the coordinator from having to do personal research or have to contact you to understand what you meant.

Here, there is no magic solution to solve this problem aside from your attention when writing your report.

ERROR 4: THE DANGERS OF "COPY / PASTE"

One of the common problems that can also be encountered in inspection reports is the unfortunate "copy / paste".

The causes of the type of error are many, but often it comes from using a previous report as a basis or copying a formula or phrase used in another section of the report.

The use of a previous report is an interesting method to save some time in drafting an inspection follow-up, but to outlaw when inspections are different. Concerning the syntax of your report, you must avoid this error by avoiding repetitions. It is useless to repeat the same thing several times and therefore the use of "copy / paste" will in fact be limited. On the other hand, you will save time writing and escape reports like "saga".

Often, the error is detected when the report is read, but this is not always the case. More seriously, this error could be discovered by your client, which will not plead in your favor and may lead to misunderstanding and approximation.

ERROR 5: THE OFF TOPIC

One of the pitfalls in which as an inspector you could fall is the non-compliance with the notification. We are not talking here about the inspection date, of course, because it is mainly a matter of organization or the address of the supplier, but the content of the mission order.

The mission order is a description of what you have to do during the inspection. The tasks to be performed are explained in the notification that you received from your client. You will see in this manual that I recommend to always do a little more than the simple respect of the order of mission but be careful not to lose sight of your first task. For example, if you are notified for a hydrostatic test do not run into coating adhesion tests. Or, at least, you can perform all the checks you deem necessary, but only after you have carried out your original inspection. So before you want to do more, be sure to respect the order of mission so as not to make off topic and above all do not forget anything.

In some cases, the order of the mission may be described fairly little. For example, the notification may simply specify "hydrostatic test". In this case, it is understood that you must do the test in question, but ask yourself what this mission implies.

In the case of our example above, this implies that a certain number of documents must be reviewed before, during or after the test. The documents to be reconsidered are deduced from the mission itself: in order to do this test, there has probably been welding and non-destructive testing. It must therefore be verified that these different steps have been carried out correctly and that the results are in line with the expected results. The subject of this manual is not to enter into this type of detail, but all this will be addressed in another manual of our collection.

In any case, it is easy to leave out some aspects of the inspection because of a poorly defined or too brief mission order. We must be vigilant and, in case of doubt, do not hesitate to contact your sponsor to let you know exactly what your mission is and what your client expects from you.

CONCLUSION

To summarize, in this first step, we were able to go through the main errors not to commit. Even if a solid report plan will allow you to avoid falling in some way, only your vigilance will be an effective safeguard to make your report a perfect working document and information.

Now that we have made the list of errors to avoid, we can deduce the 5 golden rules of the inspection report writer.

To remember :

Your report should not be a vague summary of a few words.

Your report should not be too long and overloaded with unimportant details,

avoid approximations and inaccuracies,

limit "copy / paste" to the bare minimum,

Your report must be consistent with the notification you have received.

In the next step, I would suggest a standard inspection report plan that you can use practically under all circumstances, but that you can also easily adapt to your needs.

SECTION 2: A COMPLETE REPORT

In this module you will see how to make an effective cover page, a clear and complete plan for the body of your report. I will also give you the essential information that all the coordinators expect to have a complete view of your service and react if the need arises.

We'll talk about the use of photos - we'll talk about it again in Step 5 dealing with attachments and dissemination - and finally, we'll see how to address the findings of your inspection.

The goal, once again, is to showcase your work and efficiency while passing on all the most relevant information to your customer.

We will dwell on the case of the form of report imposed by your client or even your employer. Do not forget that whatever your situation - independent or not - your customer's satisfaction remains your priority even when you have to use special shapes.

You will be able to evolve this plan according to your needs and possibly to create you basic models for each type of inspection that you will have to drive.

COVER PAGE

The cover page, although too often overlooked, is a very important part of your inspection report. It is on this page that you will both reference your report, but also ensure traceability. This cover page will also serve to ensure referencing at your client for both tracking and archiving.

It goes without saying that a minimum of information is necessary for the identification, referencing and follow-up of your inspection report, but also for the inspection itself. Here is the list that I consider to be the minimum required to ensure effective traceability for your client, but also for yourself:

✓ The number of your inspection report will be the common reference between you and your client if you have to discuss or revise it,

✓ The notification and inspection dates will not only date your inspection, but will also highlight the time between the date of notification and the date you were able to inspect your inspection. Regarding the date of notification, indicate the date of notification from the supplier to your customer. About the inspection date, you will often have to specify several days, weeks or even months. Personally, I exclude the dates of trips and returns, as these dates are only important for billing,

✓ The names and the Project number are also to be indicated to the extent that you have this information. This can be useful for locating your report in a set,

✓ The names of the client and the supplier will be used once again to identify your report,

✓ The place of inspection is also important, especially when the supplier's address on the notification and the inspection address are different. This is often the case when you are going to do inspections at a sub-supplier. Useless to indicate the complete address, city and country are sufficient in most cases,

✓ The item number or "Tag" of the material concerned by your inspection will give you a more detailed description of the subject of your report. Indicate if you know them, but in any case it should be accompanied by a short description. I recommend that you use the one indicated on the notification without interpretation so as to avoid any ambiguity,

✓ A very simple description of the inspection to be conducted will also be interesting information for your readers,

✓ The type of inspection - partial, intermediate, final ... - will tell your client whether all the equipment has been inspected or not. It is up to him to decide whether the remaining equipment or tasks should be subject to inspection or not. In this section, you can also indicate if this inspection ends control activities,

✓ The number of pages in the report including the attached documents will be of interest to your client who will know if he or she has the entire report.

One can also add optional information, but which can have their usefulness such that:

✓ The design code (EN, ASME, API ...),

✓ The applicable regulations (CE, EAC, ASME, DOSH ...),

✓ The number of the Quality plan and its sequence - you will obtain this information at the minimum when signing the Quality plan, but if the notification has been made correctly, you already have all this information.

Finally, it is very important to give the result of your inspection which can range from acceptance to rejection of equipment or testing. For my part, I use only three possible conclusions: accepted, accepted with reserve and refused. We will come back to this later, because each conclusion will have different consequences.

In order not to weigh down this cover page, I do not specify the list of attachments which will be the subject of a paragraph of your report as we will see later.

You can also contact me to receive free my form of cover page. Appointment at the end of the manual to know my details.

THE PLAN

1- LIST OF REFERENCE DOCUMENTS

This is to list all the documents used to conduct your inspection. This list can not be defined here because it depends directly on the documentation your customer has provided you and of course the type of inspection to be conducted.

Be careful, there are some important things to check before starting your inspection. Start by comparing the concordance of the revisions between the documents submitted by your client and those that the supplier owns.

In addition to comparing the revision level, be sure to check the status of each document that will serve as the basis for your inspection. It is usually considered that there are 4 statuses for a given document:

"Approved": here we consider that the document has been reviewed by the client and that the client has no comment. The inspection can be done.

Approved with comments: In this case, the client reviewed the document and made comments that may be considered minor. In this case, before proceeding with your inspection, check if the comments have been taken into account.

"Rejected": This time, the document was reviewed, but the client made some major comments and found that the document was not acceptable. If you find yourself in this situation, do not carry out the inspection without the authorization of your sponsor and be assured that the supplier has taken into account the comments in question.

"For information": The document was submitted to your client, but for his information only. Since this document is not intended for comment, you may proceed with your inspection.

If you think you do not have enough documentation, do not hesitate to ask the supplier. Do not forget to keep a copy that you will attach to your report. Also specify in this paragraph which documents were provided to you by the supplier.

2- VISUAL AND DIMENSIONAL INSPECTION

In this section, it is a matter of achieving a state based on a visual observation of the material. This is of course the first step even though it is not necessarily the simplest to achieve. This type of inspection usually gives a lot of information about the overall quality level of the equipment you are inspecting. It controls the general appearance - shock, paint, corrosion, cleanliness ... - which is a good indicator of the care taken to the material and by extension to the care taken to achieve it. Note anything you think you should report during your exam.

It is at this stage that one controls for example the temporary support, but also the traceability, or the deferrals of marking, etc.

At the end of the visual check, I do a quick dimensional check even if it's not in my mission order. In particular, check the interfaces with future connections. Once again, the accuracy of the dimensional readings is indicative of the overall quality.

It is faster to take dimensional readings directly on the plans provided by your client. If these have not been provided, it is easy to ask the supplier to print the latest revision, either overall plans or detailed plans that may be of interest to you. Be careful to ask for design plans and especially not plans as "built" because these plans have already been corrected to indicate the dimensions actually manufactured.

In the case where several parts are presented, it is possible to work on a more limited lot. Except as specifically instructed by your limited partner, good practice sets this lot at approximately 10% of the total quantity. It goes without saying that in the event of a default that you would have to note, this lot will have to be revised upward. For my part, I use the following rule: 10% - 25% - 50% - 100% the goal is obviously to have, within the time given, a most representative view of the total quantity.

Do not forget to include in your report the percentage of the lot you inspected. Once again, report the discrepancies if these are relevant.

3- NON-DESTRUCTIVE TESTING

This section of the report is important because it permits the integrity of the permanently assembled material to be checked, for example by welding or gluing. There are many non-destructive tests to highlight various defects. We will not list it here, but another handbook will be devoted to this particular topic.

Whether you attend this type of test or not, the reports will be reviewed. It is therefore important to mention this in your inspection report. Clearly indicate whether you have only done a review or if you have witnessed the tests in question. You can make a separate paragraph for each test, however, it is sometimes simpler to summarize all the tests in a table showing the percentage performed for each type of test and to indicate the final result (satisfactory vs unsatisfactory).

The review of these reports is especially important if the equipment is to undergo pressure testing, as this type of test always presents a risk for operators and for you if you are attending.

In the case of non-destructive testing, beyond the reports, it is important to mention the procedure if it exists or at least its reference number. Another point that must be included in your report is the verification of the qualification of the operators. It is therefore necessary to indicate in this paragraph that these qualifications have been checked, under what code (ASME, EN …) the operator has been qualified and his level (I, II, III). Without this check, your client would be entitled to put the tests in doubt.

In order to prove that the test reports have been reviewed, I recommend that you countersign them. We will discuss the signatures in step 7 of this manual.

4- MECHANICAL TEST AND PRESSURE TEST

In this section of the report, you will be able to discuss two topics that are often at the heart of an inspection. You will of course have to distinguish these 2 types of tests and their particularities to indicate in your inspection report. These tests, although very different, both depend on a test bench.

So to perform these tests you will need a suitable bench. Although the benches are of course different, there are still a number of points to check before starting the tests. You will have to report these checks in this paragraph of your report.

All test benches have measuring instruments. Before any test, check and report that these measuring instruments are correctly calibrated and correctly marked. If the measuring instruments are not calibrated correctly or if they are simply out of their calibration period, the test can not take place. There is a significant risk of rejection of the test, which would involve an expenditure of time and possibly money.

On the other hand, the test benches need labor: the operators. As well as non-destructive testing, operators must be qualified. All tests do not necessarily require a qualification recognized by an independent body. In this case, ask for proof of internal qualification, for example a certificate of training or aptitude.

This type of test also requires assembly and a test procedure. Do not forget to mention in your report that the assembly is in conformity and that the procedure is respected. In the event that you notice any deviations, have the assembly modified or the procedure followed, but in any case, report the differences you observed.

Finally, present the results of your tests. There are two options: either you specify all the test data, you confirm that the test is successful and you invite your reader to refer to the attachments where he will find the test report. For my part, I prefer the second solution that saves writing time. But this option is only feasible if you can put the test reports as an appendix to your report. It is not always possible. In this case, I describe in more detail the parameters of the test.

Conclude this paragraph with a conclusion indicating simply whether the test is successful or not.

5- FUNCTIONAL TESTS

Functional testing, which consists mainly of functional checking and performance measurement of a device, is a good starting point for the set of tests indicated in the procedures made available to you by your customer or by the manufacturer himself.

It is impossible to give a concrete example here, because each functional test is specific to the material concerned. But it is relatively easy to construct your table by taking the index of the test procedure or each line of the inspection plan if it is sufficiently detailed.

For my part, I use a simple table with columns: the list of tests and checks carried out, the reference to the test procedures and finally the results (compliant or non-compliant). Each line is of course a particular test or control.

Report the result of each test independently and your comments. Remember that once again there are test beds and operators, so it will be important to note in your report that you have done the verification of calibrations and qualifications.

In the case of functional tests, it is important to attach the minutes of the various tests to your report. These tests are usually quite complex and the reports are therefore often much more provided. Your customer will be interested in the precise results of the performance of the equipment he has purchased. On the other hand, some data are important for future manufacturing phases. Communicating them to your client will enable them to anticipate these next steps by exploiting the results as soon as your report is submitted and without waiting for the final documentation.

6- PACKAGING AND PRESERVATION

If you have the opportunity or if your mission statement so specifies, it is interesting to describe the type of packaging and the type of measure taken to ensure the protection of the material during its transport and storage.

There are of course several types of packaging meeting several different standards. Since I am not a specialist in this type of inspection, I use the SEI (Union of Industrial Packaging) as a reference document, describing just about all types of industrial packaging commonly used. Warning, this is not a standard, but simply an aid.

In your report, you will be able to describe the type of packaging used. It is unnecessary to describe its characteristics, but simply to verify that this packaging corresponds to the one ordered. Note in your report if you have also verified the marking.

You should also indicate the type of preservation that has been put in place. There are of course several types that can have the same result. It is therefore important to describe the method chosen. This must of course be in agreement with the method ordered.

7- THE DOCUMENTARY REVIEW

Each inspection is not limited to a physical examination but is also accompanied by a literature review. Therefore, reviewing this documentation is one of the tasks of your inspection, and you should also report what you have seen.

It is important to note in your report the list of documents that you have been asked to review. If possible and relevant, it is preferable to attach a copy of all documents reviewed in your attachments. In any case, take a copy of all the documents you have countersigned as proof of review.

The list of documents that you will need to review can not be drawn here because it depends on the type of inspection and the product inspected, but they must of course be related to your inspection. In order not to lose your reader and like, this list often corresponds to the attachments, I recommend you to list the documents reviewed at the end of the report.

We will return to these documents in more detail in Section 4 dealing with attachments.

PHOTOS AND DIAGRAMS

Photos are an important part of your inspection report. An image or a diagram often says more than a long speech. I recommend that you insert pictures or diagrams in your report. Above all, ask the supplier for permission before taking photographs of the equipment, in case of refusal, you can ask the supplier to take them himself by indicating the views you want.

For my part, I include in my reports 1 or 2 pictures of the whole product, 1 or 2 other details of the most critical parts of the product and in particular those subject to my inspection. In addition, I add as many photos as defects observed.

For detailed pictures, consider specifying the scale. This is especially important when your snapshots are used to show a defect. The best thing is to add a dimension, but you can also place next to the subject of your photo a sufficiently common object to allow the reader to deduce a dimension such as a pen.

Always take color pictures with a good definition. It is possible that your pictures are reduced to a vignette, so it is very important to have a sufficient level of definition of the photo so that the images remain clear. Do not add black and white photos to your report, as the images will be much less clear, especially in case of reproduction.

The layout can be done in different ways: either at the end of the report or in each paragraph. Personally, the option "end of paragraph" is the one I choose commonly for my own reports. Whichever layout you choose, you must always attach to each photo a legend sufficiently clear for the understanding of your reader. This should be short, but sufficiently clear to easily identify what the image represents.

You can make videos if this is useful, but remember that this file should be sent separately from your report. On the other hand, videos are often too large for an easily transferable inspection report. Make videos only if it's really relevant and important for your client to really use.

CONCLUSIONS

An inspection report must also have a clear and unambiguous conclusion. "Accepted", "accepted with remark" or "rejected" are the three possible opinions you can make. Your conclusion can cover all the inspected material or you can detail it for each item or lot when they do not all have the same final status.

I recommend that you indicate your conclusion on the cover page, but also at the end of the report. Feel free to enlarge the font size to make your conclusion stand out from the rest of the report. Your reader will have the conclusion of your inspection without having to read your entire report immediately.

This conclusion is important for your customer, but also for the inspected supplier. It is your conclusion that will give the status of the material.

You must also be careful that your conclusions are not influenced by elements outside your inspection. It goes without saying that you should not give in to any form of corruption, nor should you be influenced by a sometimes difficult relationship with the supplier. In any case, your conclusions must be objective and factual.

THE USE OF AN IMPOSED FORM

Sometimes your client already has a particular form for his reports. Use it taking care to perfectly fill the shape in question. Generally, this form is limited to a particular cover page, but leaves some freedom as to the writing of the body of the report. It is therefore in this part that you can develop the plan proposed above.

When you use a mandatory form, most of the information you will need to complete your report will appear on the inspection notification. However, it is possible that a certain amount of information to indicate, especially on the cover page, have not been communicated to you. In this case, do not hesitate to contact your customer to obtain them. If you can not reach your sponsor, tell them that there is a missing information on the cover page in your broadcast message.

CONCLUSION

You now have all the information you need to compile a complete inspection report. This plan that I have been using myself for several years has never been the subject of negative comments.

Remember that you are the representative of your sponsor and as such you must provide a full report with a clear plan and clear conclusions. Do not underestimate the importance of the cover page, but also the accuracy of the information given.

To remember :

The cover page must be carefully filled.

The plan must be clear and address all the topics of your inspection.

Images are often more meaningful than long speeches.

Every inspection must have a conclusion.

Although this type of report is already more than enough to satisfy your client, you can go further and stand out from other inspectors. This is what we will see in the next step: little more.

SECTION 3: SMALL

You've seen how to write an effective and comprehensive inspection report, but is it really enough to stand out and be recognized as an inspector whose reliable reports are valuable supports for customers always looking for quality? Can you do a little bit more without compromising your inspection and the quality of the report?

For my part, I recommend you add a few paragraphs that will prove useful to the reader even if this is outside the scope of your original mission statement. Sections such as order progress and security are important information that will further enhance your report and thus your reputation.

Of course, that's not all, when you release your report, you can obviously share your recommendations and convey relevant information without being included in your report. You will also be able to communicate this information if, like me, you are in the habit of sending "Flash messages".

COMMENT ON THE PROGRESS

Whatever type of inspection you are required to do, intermediate or final, a progress point can be useful information for your client. Indeed, it is always interesting for an inspection coordinator to know, in addition to the "remain to be done", the confirmation of the key dates and therefore of the planning. The contractual aspect of respecting the manufacturing and delivery schedule remains, whatever the result of your inspection is particularly useful for your customer. It remains to be seen what the important dates are.

In reality, there are very few really important dates to have a clear picture of the progress and the respect of the contractual dates. Keep in mind that normally your customer already has the full manufacturing schedule for the material you are inspecting, so it is not necessary to take stock of all the planning lines. So what are the key dates you can ask for without losing time during your inspection and giving your client a clear idea of the situation?

The dates you will have to check and indicate in your report are as follows:

- Dates of the next tests,

- Date of final inspection,

- Date of packing,

- Delivery date.

Of course, depending on the time of your inspection, certain dates will no longer be current and therefore you will obviously not have to indicate them in your report. Each of these dates represents a key step in the manufacturing process.

The dates of the next tests will inform your customer not only on the progress of the manufacturing process, but also to anticipate future inspections. Your customer will therefore be able to anticipate the inspections and to plan the inspectors who will carry them out. It will also be able to manage its consumption of hours and its inspection budget.

The final inspection date is an important milestone as it concludes the manufacturing phase. This is an important date from a contractual point of view which makes it practically possible to determine the date of packaging and delivery. It is nevertheless better to communicate the latter.

The date of packing also makes sense in particular when it comes to packaging. It may be interesting to indicate whether this task is subcontracted or not. As a reminder, each subcontracted operation requires more vigilance, as the advancement no longer depends only on the official supplier, but also on a third party over which the customer has little control.

The delivery date is the most crucial date for your customer. Be careful, however, because this date can be considered in different ways depending on the "Incoterm" contract. There can be a big difference between the so-called "ex-Works" date and the "delivered site" date especially if the site is at the other end of the world. In this case, it's up to you to choose which date seems the most relevant. For my part, most often, I indicate the date "ex-Works" which seems to me the most useful especially to proceed the "release" of the material, but also to make the request for the final documentation. This date is also important to plan a particular transport if it is an exceptional transport.

SAFETY COMMENTARY

This point is never on the agenda of your inspections and yet it is a point that concerns us all. I like to say that we practice our profession to earn a living and not to lose it. Security is everyone's business and in all circumstances. Certainly, the terms "safety first" are in fashion in recent years, but they are not just empty words.

An update on security in your report will demonstrate your involvement in this joint action. So do not just put a paragraph concerning only your own security, but also security in general. For my part, I observe the workforce as the material at the same time as I conduct my inspection. If you do the same, summarize in your report for both the supplier and your client. Security has many human and financial impacts on manufacturing, so making it a separate section of your report is fully justified.

Regarding this point in particular, ethics requires you to be vigilant for yourself, but also for all staff working on the site and likely to be put in a dangerous situation. It is your duty to alert your interlocutors if you witness a risk situation. In addition, you also have the duty to stop all activities that may be at significant risk, but be reassured, no one will hold you, quite the contrary.

Of course, you should be concerned about your own safety at all times, especially at the time of inspection. As soon as you access the site, you must be vigilant, because you enter a workspace potentially involving risks specific to the activity of the site. Begin by having you explain the safety rules. Access conditions, individual protection, evacuation plan and emergency number are the minimum information that you should ask your contact if it does not provide you spontaneously. If he does not, do not hesitate to write it down in your report and remind him.

Wear your personal protections, if you do not have them, ask your contact. I remind you that it is his responsibility to receive you safely. If it is not able to provide you with them, you should also note it in your report and, if justified, use your right to withdraw. In extreme cases, if you feel that your safety is not assured, you have every right to invoke this right of withdrawal which allows you to end the inspection. Of course, this action must be justified and justified.

During your inspection, check the procedures for carrying out the tests, check the machines used in your activity, and also check the competence of the operators who assist you or who are conducting the tests. Also check your environment to make sure the test can be conducted and in case of accident all conservative measures are expected. For example, if you have to drive a test involving nitrogen, you should not be in confined space unless you have adequate safety equipment.

During the inspection, do not hesitate to observe around you and note all that seems to you to be dangerous as much for all that material - poorly used bridge crane, blocked passage way, risk of falling object ... as regards the human being -

inadequate behavior, lack of personal protection, incorrect use of equipment ... -.

At the end of the report, indicate by a short note what you think is important to report. Keep in mind that your feedback is as much about your customer as it is about the vendor, but also that it may have a longer-term impact than just your inspection.

This subject is very important and very broad, and will be the subject of a specific guide.

ADDITIONAL INFORMATION

The additional information must not appear on the inspection report, it is not mandatory and you may not have it. Anyway, if you have any, send them by e-mail or orally.

The additional information can be of any order. Regarding the "follow-up" for example, if you see a busy workshop - you can deduce a possible risk of late delivery - or an empty workshop - which may be a sign of a decline in load endangering the durability of the company - or if you find the quality poor even if it remains consistent, etc. Attention, the information that you transmit must be factual and motivated.

For my part, I share particular information - when I have course - by e-mail when submitting my report to my client. However, it is rare for me to transmit this type of information except for long-term clients who ask me for it.

You can also send additional organizational information if you think it is relevant especially to anticipate certain events such as loading the material or the importance of the number of packages for shipping.

Other cross-sectional information may be important. For example, you can inform your sponsor about a loss of certification, a scheduled shutdown of production or information about the apparent health of the company. The risks of redemption or closure outright are possible. It goes without saying that your customer will be able to use this information to guard against any change in the functioning or the longevity of his supplier.

In short, the information can be of any nature and have the purpose of being communicated to your customer. Be careful, however, to ensure your duty of confidentiality.

THE FLASH MESSAGE

Here is another process that customers generally like: what I would call the "Flash" message.

I practice it myself when I do an inspection, but I also appreciate receiving one when I am in the position of the inspection coordinator. The principle of the "Flash" message is simple: send a short e-mail message to your customer while you are still at the inspection site but your inspection is complete.

What should this message contain? Simply the result of your inspection without giving more details that will anyway be available in your future inspection report. Also state the exact or approximate date of your final report.

Remember that the message is meant to inform your client of the status of your inspection, so you must make a short but straightforward message: "accepted", "accepted with comment", "rejected" . In the last 2 cases, state succinctly the reason (s) for your conclusion.

Depending on the severity of an inconclusive inspection, indicate only what is problematic so that your client can anticipate your report and start conducting the actions that it deems necessary.

CONCLUSION

You now have some additional keys to distinguish yourself from the other inspectors, but also, and above all to show a professionalism that customers particularly appreciate. It's these little extras that will help make you an indispensable inspector.

Be careful, always remember that you are bound to a duty of confidentiality? This notion is particularly important especially when you have to communicate particular information. Always keep in mind the interests of your client.

To remember :

Make a progress report on the order.

Report the level of security.

Submit additional information.

Send "Flash" messages whenever possible.

SECTION 4: THE PUNCH LIST

When your inspection is not conclusive or, in other words, the material is subject to comments, then your report must be completed with a punch list. The inspection report reports a finding, but the "punch list" is a list of actions.

The "punch list" lists all the operations still to be performed and the corrections to be made to ensure that the equipment is in full compliance with the specifications that your client has provided. This list should only contain points that make the material potentially non-compliant.

We can summarize this list in 3 particular points: What? Who ? When?

The what ? Does not define the defect or the incomplete task, but sets the action to carry out as precisely as possible to resolve the defect. Since the definition of the defect is already in your report, you must specify a short description focusing on what needs to be done. This description should be simple, but clear enough to be understandable to all parties. Do not forget to refer to the relevant paragraphs of your report.

Who ? Defines who will be in charge of carrying out the stated task. Attention it can be the supplier, but also the customer. Responsibility is defined according to the task at hand. There must be only one responsible to avoid any misunderstanding between the parties. Avoid naming a person's name, but instead indicate the name of the company responsible for carrying out this action.

When? Is always a date, but it can be understood in several ways: as the deadline for execution or as the actual date of execution of the task. Knowing that this doubt is possible it is important to indicate on the punch list which type of date is indicated. This date does not have to be an estimate in the form of a week or a month, but it must be indicated as "day / month / year" or equivalent.

THE FORM

The form remains relatively simple, but you can also develop more complex forms that are better adapted to your needs or those of your customers. Your customers may also tell you their own forms.

The basic form is a simple table on three columns: the what?, The who? and when? In most cases, this simple form is sufficient. As indicated in the previous paragraph, it is possible to elaborate more complex forms in each of the parts while keeping the basic principle of "What, Who, When".

The what ? Is intended to further detail the task to be performed by indicating the task itself. Again, you can add fields such as the qualifications required to perform it, the way to verify the execution or the quality of the execution or the contractual or regulatory related tasks, etc. Do not add unnecessary fields such as the causes of the problem to avoid any form of repetition with your inspection report.

For the "Who? In addition to just indicating the name of the company responsible for the task, you can specify the name of a contact, but in an informative manner only. In case of change of contact, the identification of the "who" of the task will remain valid. You may also, if necessary, indicate a contact address, a geographical indication, etc. Be careful once again to ensure that this additional data only concerns the identification of the person responsible for the task described.

For the "When? You will have to decline the start and end dates, the actual duration of the transaction or other dates such as the date of receipt of proof. These dates must be precise and in identical form. Avoid dates in the form of weeks, months, or years.

We can also add columns to indicate information such as among others, the impacts planning, costs, etc. Attention it is important not to unnecessarily overload the "punch list" with information that has no direct interest in closing it.

PHOTOGRAPHS AND OTHER MEDIA

The punch list can be embellished with photos or other media such as videos or diagrams. All forms of explanation that can ensure the clarity of your words are welcome.

Since, at the time of writing, the tasks have not yet been completed, it is difficult to add any media to your punch list. You must therefore carefully choose the parts you will bring.

You can refer to the documents and media that will be part of the attachments. As we will see in the next chapter - where we will talk more about photos - with a good referencing of your attachments, it will be easy to refer to them from your "punch list".

RESPONSIBILITIES

As in all actions requested for correction or improvement of quality, you will have to appoint a person in charge. This responsibility, defined by the task itself, will be given to a company or a natural person responsible for the proper execution of the task.

There may be several officers, in this case, preferably appoint the company - or the person if necessary - to coordinate the actions defined beforehand and to ensure that they are properly carried out.

The tasks of this manager will be to:

- validate that the task has been performed,

- communicate the completion of the task,

- provide proof of performance of the task,

- close the task.

The manager is therefore the focal point of the closing of the punch list, so it is important to select it carefully, according to the task, and to indicate all the information necessary for good communication. The latter must, of course, validate the "punch list" from its inception in order to remove all forms of ambiguity in the actions to be carried out.

MANAGING THE PUNCH LIST

As an inspector appointed for a single inspection, you will not be in charge of complete manufacturing follow-up. The responsibility for verifying the correct closing of the punch list is therefore not yours. This charge will accrue to your customer.

On the other hand, if you are mandated to follow the manufacture to its end, it is your responsibility to validate the closure of each point of the punch list. It is therefore possible that, in a follow-up report, you have to add an old punch list to report on its progress. In this case, do not forget to recall in your report the references of the original report from which the punch list is based. Again, there may be several, so it will be important to be precise in the referencing of the punch list and in the closing documents accompanying them.

It is also possible that you have to manage the punch list during the inspection. This is often the case when your mission lasts several days. In this case, whether you are the customer or simply the inspector, you will have to evolve the "punch list" throughout the inspection by closing the open points that will be solved at the end of the inspection. your mission.

Attention, depending on the type of task, the closing of the task can be the subject of a new inspection report and therefore it is still possible that you have to draw up a new "punch list".

CONCLUSION

The inspection report states findings contrary to the "punch list" whose role is to define the actions to be taken to correct or improve the equipment inspected. In this section, you have seen why and how to write a punch list that can be easily used for any speaker in each party.

Remember to make it as simple as possible and avoid overburdening this document unnecessarily, especially with attachments that will in any case be grouped together in another part of the inspection report.

To remember :

Make a list in the most appropriate form.

Insert images when necessary.

A punch list must be followed and the tasks closed.

SECTION 5: ATTACHMENTS

An inspection report is usually accompanied by attachments. These documents contain all the documents that will be attached to your inspection report to support it. The attachments can be of different natures and have the function of providing the proofs of the tests and checks carried out. Most of the time, these documents are documentary forms, for example test reports or material certificates, but they can also be photographs, videos or other media (radio film, sound recording, graphics, etc.).

PHOTOS

One of the most common attachments is photography. They may appear in your report or be attached to it. In any case, the photos must be particularly well referenced and succinctly described.

Photos must be sharp and colorful, but sufficiently clear to be printed or reproduced in black and white.

Photos can be transmitted as a thumbnail in the report. In this case, I recommend to also transmit the native files of the photos so that your readers can have access to an optimal quality. They can then make enlargements or other transformations such as, for example, framing or format.

There are 3 main formats for digital photos: ".jpeg" or ".jpg", ".gif" and ".png".

JPEG is the most known and used format. This is the best quality / weight ratio of the file with its 16 million colors. This is the preferred format for an effective inspection report.

GIF is also a fairly common format with which you can manage transparency, but offers a color display limited to 256, but their weight is very light. This format is poorly suited to the inspection report by its lack of image accuracy.

The PNG also offers the possibility of managing transparency, but with the 16 million colors offered by JPEG. Unfortunately, the files are very heavy and therefore not very suitable for inspection reports.

In addition to the file format, the size of the photos is important. It mostly presents itself as a pixel such as 700 x 500 pixels. Generally, the size at the output of your device, whether a camera or a smartphone, is usually much larger than the proposed example. It's up to you to adjust the size to fit your report so you do not overweight your report.

Finally, there is also the resolution to be taken into account. I recommend you limit yourself to the maximum that can display a screen: 72 pixels per inch. Do not go below this value or you will lose quality. More pixels per inch would be unnecessary as part of an inspection report.

For the record, you can consider that if your image is still close to 1 MB then your image can be further optimized. To do this, consider using thumbnail. For example, an image at 300 dpi at a weight of about 4 MB while a 72 dpi image is only 250 KB. On a thumbnail, the difference can not be made with the naked eye.

I recommend that you use the thumbnails as a board and as proposed in a previous chapter to provide the photos in their original format in a separate compressed folder of your report.

REPORTS AND CERTIFICATES

When you attend any tests, one or more reports must be issued by the person or company in charge of these tests. These reports are official documents, often contractual and must therefore be included in the attachments.

You must countersign each report or certificate and more broadly each attached document proving that you have carried out the documentary review. In addition to the words "Approved", indicate the date, also to appear on these attachments. You must sign the originals of the supplier and attach a copy to your report.

Be careful, when a document is rejected, it is particularly important to attach it to the report to justify the rejection, but also because it will trigger a series of corrective actions. This document should certainly be referenced in the "punch list" and possibly in a non-conformance.

ANNOTATED PLANS

Attachments may also include annotated plans such as "as-built" plans or plans showing non-conforming points. These plans may also be schematic or graphic.

This type of attachment is particularly useful for illustrating your purpose, especially when you need to refer to specific defects in particular locations. The defect can thus be precisely localized and possibly described. As is often said, an image is much more meaningful than a long speech so do not hesitate to use this method to detail your inspection report.

Attention, your notes should be clear, do not hesitate for it to include only a particular view of the plan. Avoid, as far as possible, color schemes or diagrams that will disappear when printing, but especially during copies.

OTHERS (VIDEO, FILES)

As stated above, there are many possibilities of attachments such as, among others, videos, radiographic films or audio recordings. Do not hesitate to forward them if you think that this type of document can provide relevant information to support your report.

Be careful not to overload this section of the report with an excess of sometimes useless documents that could interfere with the understanding of your report. It is, for example, useless to make a video for a hydrostatic test lasting several hours or even a few minutes. On the other hand, it may be interesting to make one to show a gas leak during a pneumatic test.

Often, the test reports are largely sufficient and allow a perfect understanding of your report and the tests you have performed.

REFERENCING ATTACHMENTS

In order to find the information cited in your report, it is essential to make references to the documents you have attached. For these references to be clearly understandable, you must reference your attachments individually and specify a number of pages. For example, you can choose a form such as "Appendix 1 - n pages".

There is no need to specify a document title. This title will be in the body of your report and most of the time on the document itself.

You must list the attachments in your inspection report, indicating of course the references you assign to them, as well as the title of the document and the corresponding number of pages. The number of pages is useful to ensure the reader owns all the pages. It may happen that some pages are missing when printed or printed. Your report must remain complete.

A good referencing of the attachments will allow you to shorten your report or to support your statement more effectively and more precisely.

DISTRIBUTION FORM OF ATTACHMENTS

As their names indicate, attachments should be posted as an appendix to your inspection report.

The best method is to make a single .pdf file containing both the report, the punch list, if any, and the attachments. This format reduces the weight of your file and facilitates its transmission electronically.

To ensure the proper exploitation of these parts, it is sometimes recommended to distribute the attachments in their original form. This is particularly the case for photos, but also plans.

The advantage of distributing this type of attachment in their native version is to let your reader exploit them according to his needs. Photos, for example, can be enlarged or used by your client in one of his presentations or his own reports. The plans for their part are very rarely in A4 format. A broadcast in this format can make it potentially unreadable: an A1 format is perfectly unusable when it is reduced to A4 size.

Whenever possible, "lock" documents in .pdf format to avoid any form of modification. Most .pdf applications provide this type of capability.

Remember to keep a copy of the original version in case of dispute. You may need to provide your inspection documents later.

CONCLUSION

We will see in a future chapter that the attachments will go a long way to saving you time. But the interest of a well-completed report of attachments, will raise the overall quality of the report and also will be a plus for your customer but also for the supplier.

With the attachments, your report will thus constitute a complete file of the manufacture at the moment of your inspection.

To remember :

A report must be supported with attachments.

Attention to the format of the photos.

A photo must always be well referenced.

The form of diffusion must be reflected.

SECTION 6: SAVE TIME

Writing an inspection report is an operation that can be relatively lengthy if you want to make a complete and well-documented report. Unfortunately, there is no quick way to write a report in minutes, but there are several ways to save time.

You will see in this chapter how you can save a little time in the writing exercise.

WRITE ON THE SPOT

One of the most effective ways to save time is to write your report on the site of your inspection and when possible as you go. There are often many downtimes during inspections, so you need to take advantage of this to start writing your report.

Some of the questions that you will discuss with the supplier will be done in an office, you can write down the answers that will be provided directly in the body of your report. Instead of transcribing your answers on a notebook and copying them into an electronic format, you should write your report directly whenever possible.

When writing your report directly to the supplier, do not worry about the order of writing. It does not matter if you do not write your report in order, you will have plenty of time to organize your report later by rewording your paragraphs with word processing.

Of course, this method is effective only if you have prepared your report in advance in accordance with the charges indicated on your notification. The plan and reference documents are, for example, sections that you can prepare before your inspection and that you will only have to update with the details of your mission and the additional information that you think fit to add .

ANNOTATED DOCUMENTS

A good way to save time is to use the documents contained in the inspection file. Your customer has provided you with reference material, for example, plans, inspection plan, specifications, or procedures ... as many documents as you can annotate directly at each point you control. The other advantage to follow these documents is to use them as a "checklist" and to guarantee that you have not forgotten anything.

Similarly, you can annotate part of the document and extract it from the document and paste it directly into the body of your report. Thus you reduce your writing time by copying text quickly without making any mistakes in interpretation. For example, instead of copying the list of reference documents you can make a screen print and copy the whole into the corresponding paragraph.

Personally, I use this method in the valve inspection framework by using the specifications as a control check list and adding a column indicating my comments. It should not be abused, of course, but this method is quite effective both to save valuable time and to produce more accurate reports.

THE USE OF "COPY / PASTE"

In section 1, we have seen the risks of misusing this method, however it is sometimes interesting to make use of copy / paste but we must be particularly vigilant when writing and reviewing the report.

There are several possible situations, one of which is to take back a report already written in the past and adapt it to your current inspection. Of course for this to be effective, the model report must be the same type of equipment. It is strongly discouraged, for example, to use an electric motor test report to make it a pressure test of a tank.

Be careful to read the report carefully to avoid leaving information from the original report. Mistakes can have significant consequences in addition to damaging your reputation as an inspector.

Another method is to copy passages from the reference documents. No need, for example, to completely copy a list of valve tags, just make a copy of the list that was given to you for your inspection and copy it in your report. As we saw in the previous section, if you do not have the native format of this list you can make an extraction or a screenshot. You can have the same approach with plans or schemes such as PID.

Remember to carefully re-read your report to ensure that no erroneous information is included because of excessive "copy / paste".

USE OF REFERENCE DOCUMENTS

An important source of time saving is the use of reference documents. In the first paragraph of your report, you have already made a list of all reference documents, so you can use them instead of long repetitions of text. Simply recalling the number or title of the document in question is often more effective than an interpretation of it. On the other hand, documents are mostly contractual, the use of reference documents avoids all forms of dispute between the supplier and the customer.

For example, during a hydrostatic test, simply indicate the start and end times of the test (or duration), the test pressure and the manometer number. Add that the entire test was run using procedure # xxxxx. In this case, it is useless to recall the position of the manometers, the details of the assembly, etc. A certain amount of this information will be included in the reports that you will add to yours.

The use of the reference to the procedure or to a specification takes all its importance in the case where the documentary mass is very important.

Take the case of a valve specification. It would be unnecessary to repeat all the data characterizing the valve if you refer directly to the number of its specification.

CONCLUSION

As you will understand, in the context of customer satisfaction, a rapid transmission of the report becomes very important. Therefore, save time in writing without omitting anything and with all the precision required is paramount. This becomes even more important when you have to chain several inspections as a result.

For your own organization, I recommend that you anticipate the documents that may serve as a reference when preparing your inspection. This subject is, moreover, dealt with in another volume of this series.

To remember :

Write report on site.

Comment directly on inspection documents.

Copy / paste but carefully.

Use the reference documents.

SECTION 7: DISTRIBUTION OF THE REPORT

At this point, your inspection is complete and your report is ready and ready for distribution to your client and supplier.

However it is possible that in intermediate phases, you have to transmit "flash reports" to inform your customer as a long inspection.

The transmission of this report must be done in the most efficient way, in an adequate format and certified by your signature.

THE FORMAT OF YOUR REPORT

The format of your report is of course important for it to be readable and usable by your customer, but also by the supplier. You must also look after the computer format of the report that you have written as part of your inspection mission.

One of the most common formats for document and readability by all is the .pdf format. Today, all word processing or spreadsheet applications offer the registration of documents written in this format. I highly recommend this format especially for its ease of use and speed of implementation.

The use of this format has many advantages:

You can collect your report and attachments,

If your report has many documents, you can create navigation tabs,

You can lock the document and avoid all forms of modification by a third party,

The weight of the file will be considerably reduced.

However, the .pdf format also has disadvantages: the documents transmitted in this format are difficult to use as a working document by your client. Especially when they are locked in read only mode - which is always preferable for the report itself - and therefore unmodifiable which could be annoying for some documents making up your report.

Indeed, certain documents must be able to remain usable: the "punch list" in particular, but also the photos. You must therefore provide a different format for all documents that may be used by either party.

I recommend that you proceed as follows:

Send the full report in .pdf format.

In a folder (compressed if necessary), pass the "punch list" in its native format.

In a folder (also compressed), send the photos in their original size and size.

With this system of files and folders, you will be able to transmit not only a fully compiled report, but also all the parts you deem useful in their exploitable version.

DISTRIBUTION TO THE CUSTOMER

After the compilation of your files in the desired format, it must be distributed to your client.

Your client is waiting for your report, so it is important to send it to him as soon as possible. The best is to transmit it the same day, but unfortunately it is rarely possible for a simple question of availability. The inspection usually takes all the time you are given and you are forced to write your report when you return to the office or in transport.

The average "acceptable" time for a customer is usually 48 hours. Set this limit, especially when there is a "punch list" of which you are not the administrator. In this case, always try to set a limit of 24 hours to allow the various parties in action to begin closing the outstanding points. It is often difficult to meet these objectives, so in some cases you can only broadcast part of the report and, as we will see in the next paragraph, you may be forced to send a "flash report".

Once your report is compiled, you can send it by e-mail if the weight of the file or files allow it, by a "large file transfer" such as Google Drive if the weight is too important. If you send your report by LFT, do not forget to send a short message to your customer to inform them of this method of transmission.

Some customers have at their disposal an LFT that they make available when needed.

Also ensure that the report is received correctly. To do so, a simple adjustment of your e-mail application can allow you to receive acknowledgments. This will give you confirmation of the correct transmission of your report.

Once you've released your report, always consider archiving a copy for yourself in each of the formats you've submitted (.pdf and native versions).

THE "FLASH" REPORT

Inspections may take place over several days or even over longer periods. In this case, you must keep your client informed of the progress of your inspection as it unfolds. Especially in the event that you detect noticeable anomalies.

It is not necessary to write a complete report whenever you want to inform your sponsor. As the name suggests, the report "flash" is a very short report containing only the major information summarized in the form of a short message possibly accompanied by one or two attachments. Especially if you have opened a "punch list" that you can make live throughout your inspection.

You can send this message by e-mail at the end of each inspection day or at a frequency previously heard. In any case, even if your customer did not make the request, I recommend that you transmit these "flash" reports which will always be very appreciated.

The "flash" report is also valuable in case you are unable to report quickly. In this case, your customer will have at least a minimum of information allowing him, if necessary, to take the necessary actions for the smooth running of the order.

DISTRIBUTION TO THE VENDOR

Once your report is written and distributed to your client, the supplier involved in your inspection must also receive a copy of all documents that you have already sent to your sponsor. Keep in mind that your customer and vendor will be in direct contact to resolve any problems you have detected or simply to confirm acceptance of the hardware or tests. It is therefore essential that the "customer" report and the "supplier" report be perfectly identical.

For the distribution itself, there are two scenarios: either you are the customer or you have been notified.

If you are the client yourself, it will be your role to convey the report. You will have to send this report either by e-mail or by LFT according to the weight of the file. In this case, you can only send a single .pdf file that includes the entire report and attachments. If you have issued a "punch list", I recommend that you retain ownership of the native version of the document, but send it separately from the report to facilitate its administration. As always, make sure that your documents are received properly.

If you have been sponsored to do the inspection then your task is much simpler: your client will handle the transmission of the report. So you do not have to worry about it. The follow-up to be given does not belong to you and you should not be involved in the distribution of the inspection documents.

THE SIGNATURES

Your report must be signed.

Your signature can have several formats: the manual signature, the signature "scanned" or the electronic signature. It is of course possible that your report is signed in a format, but also be composed of documents whose signatures are in a different format. For my part, I mix formats to get a report properly and fully signed.

The inspection report is signed by electronic signature at the time of its writing or conversion into a .pdf file. Most word processing or .pdf applications offer this function. These applications offer templates of signatures indicating various information. For a report signature, the following information must be kept: name, first name, date, name of your company - if there is one, which is not always the case for the self-employed. There is no need to retain information such as the time or geographic location that is sometimes displayed.

This will save you from having to print and scan your document. Do not forget that saving time is one of the key keys to delivering your report on time.

As we saw earlier, the attachments - test report, certificates ... - are normally signed on the spot. Therefore, use your manual signature. As for the date and your name, I recommend investing in a custom date stamp including your name, company name and date. You can find them on Amazon for example or on more specialized sites for a price ranging from about 15 to 50 €. Thanks to this type of stamp, you will save valuable time and just add your parafe.

The photos are not signed, but you must add a caption and a date on all photos or pictures - diagrams, plans, etc. - in your report.

CONCLUSION

In this chapter, we covered all important aspects of the distribution of your inspection report. Keep in mind that often the container is as important as the contents, and your report should be disseminated in the shortest possible time without exceeding the 48-hour limit.

To remember :

Attention to file format.

Beware of broadcasts to the sponsor and the provider.

All documents must be signed.

LAST CONCLUSION

Here we are at the end of this manual. We went through the most important aspects: the mistakes to avoid, the complete plan of a good inspection report, the principles of the punch list, the attachments.

We've also touched on other things that are sometimes overlooked, such as time-saving methods, small details that can make your report a more interesting document for your client, but also the best way to disseminate your report.

Remember that your internal or external customers are still waiting a lot for an inspection report because they consider you to be their eyes and ears. They depend on your report to make decisions that can sometimes have significant impacts on the smooth running of their project.

The quality of your report is therefore paramount. Treat the content, but also the form.

Finally, you must do all this work in record time to ensure proper tracking especially when there are "punch lists" or nonconformities.

All these recommendations are only intended to provide you with a method to increase the quality of your inspection reports and, by extension, your reputation as an inspector. An effective inspection report will bring you the extra reputation you need to establish a relationship of trust between you and your customers.

###

NOTICE AND CONTACT

I thank you for reading this manual and hope that it has served you in your activity as inspector, coordinator or even Quality Control department manager. Feel free to leave your comments on its sales page. I am keen to consider your comments and make any necessary changes.

This Manual is part of a series dealing with Quality Control and Coordination. Other volumes are in preparation such as the preparation of an inspection, the "punch list" in detail, nonconformities, pressure tests, etc.

Whether you have effective methods or topics that you would like to see, if you wish to have other information or share your experience do not hesitate to contact me by e-mail: mailto:l.gaillard@qcleaks.com

If you want to follow our collection, I invite you to register on our website at the following address: qcleaks.com

THANKS

Many thanks to my review committee for their recommendations and corrections.

Thank you especially to my lovely wife who had the patience to support me during the development of this manual.

Finally, thanks to Colbert without whom I might be unemployed.

"If our factories force the superior quality of our products by force of care, foreigners will find it advantageous to supply themselves in France and their money will flow into the coffers of the Kingdom."

Colbert - August 3, 1664

THE EFFECTIVE INSPECTION REPORT

Collection QCLeaks

Ludovic Gaillard

Cover: © Ludovic GAILLARD

All Rights Reserved

ISBN 13: 978-1979019965
ISBN 10 : 1979019967

Published by QCLeaks,
Roissy en Brie, France

www.ingramcontent.com/pod-product-compliance
Lightning Source LLC
Chambersburg PA
CBHW040231220526

45473CB00001B/197